My First Science Books
A CRABTREE SEEDLINGS BOOK

FOOD CHAIN IN A DESERT

Alan Walker

A desert **habitat** has very little water.

4

Yet many plants and animals live there.

Jackrabbits dine on the desert plants.

Jackrabbits are **herbivores**.

Jackrabbits make a tasty meal for rattlesnakes and other **carnivores**.

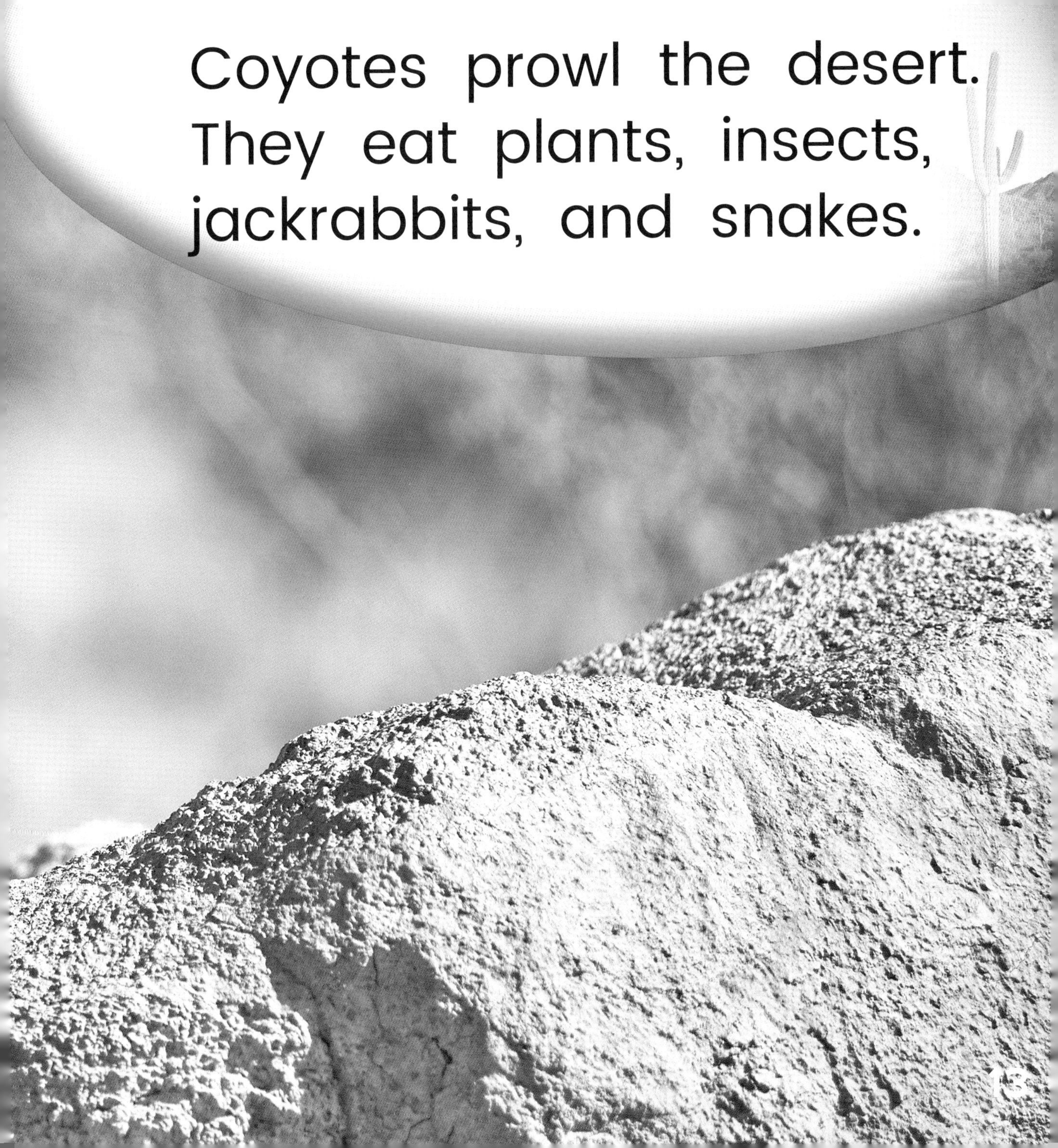

Coyotes prowl the desert. They eat plants, insects, jackrabbits, and snakes.

Coyotes are **omnivores**.

Vultures clean up habitats. They eat **carrion**.

Vultures are **scavengers**.

This is a desert food chain.

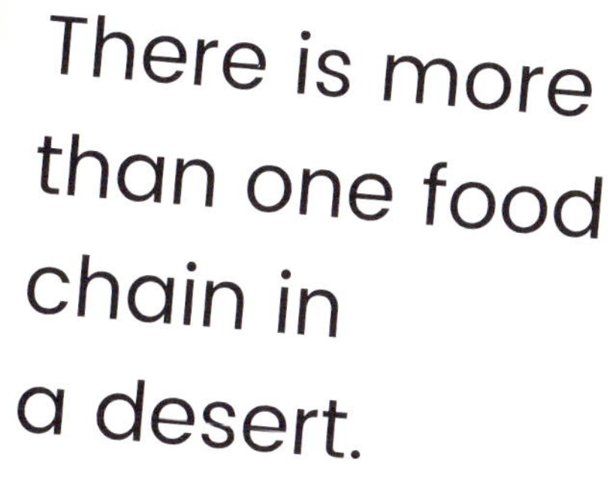

There is more than one food chain in a desert.

Read about other desert animals. Draw a different food chain.

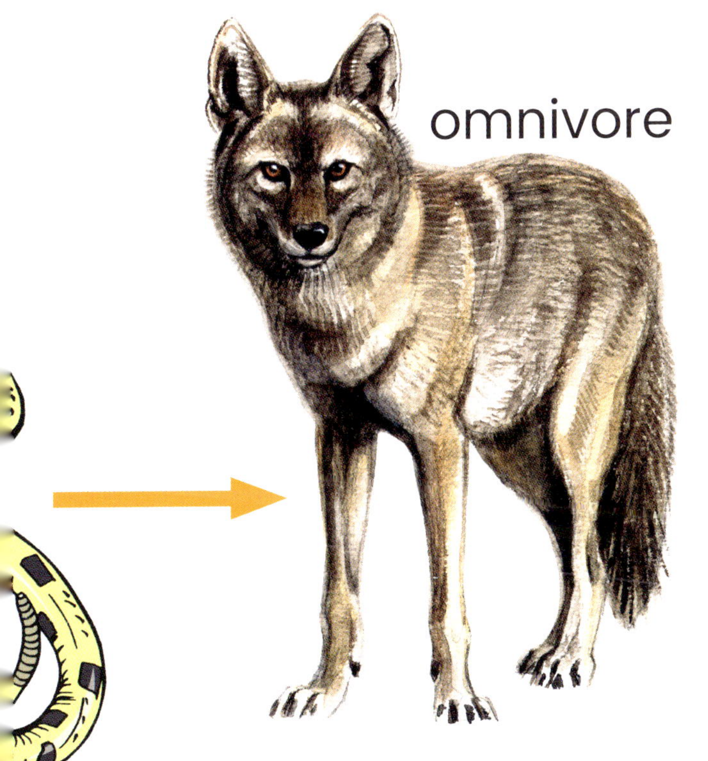

Glossary

carnivores (KAR-nuh-vorz): Carnivores are animals that eat only other animals.

carrion (KAIR-ee-uhn): Carrion is the rotting flesh of a dead animal.

habitat (HAB-uh-tat): A habitat is a place a plant or animal naturally lives.

herbivores (HUR-buh-vorz): Herbivores are animals that eat only plants.

omnivores (OM-nuh-vorz): Omnivores are animals that eat both plants and other animals.

scavengers (SKAV-uhn-jurz): Scavengers are animals that eat carrion.

Index

School-to-Home Support for Caregivers and Teachers

Crabtree Seedlings books help children grow by letting them practice reading. Here are a few guiding questions to help the reader with building his or her comprehension skills. Possible answers are included.

Before Reading

- **What do I think this book is about?** I think this book is about a desert food chain. It might tell us about the animals that are part of a food chain.
- **What do I want to learn about this topic?** I wonder which animals are parts of a desert food chain. I see pictures of four animals on the cover.

During Reading

- **I wonder why...** I wonder why animals live in deserts where there is very little water.
- **What have I learned so far?** I have learned that jackrabbits, rattlesnakes, coyotes, and vultures are part of a desert food chain.

After Reading

- **What details did I learn about this topic?** I learned that plants are at the beginning of a desert food chain.
- **Read the book again and look for the vocabulary words.** I see the word ***omnivores*** on page 14 and the word ***carrion*** on page 17. The other vocabulary words are found on pages 22 and 23.

Library and Archives Canada Cataloguing in Publication

Title: Food chain in a desert / Alan Walker.
Names: Walker, Alan, 1963- author.
Description: Series statement: My first science books | "A Crabtree seedlings book". | Includes index. | Previously published in electronic format by Blue Door Education in 2020.
Identifiers: Canadiana 20200389548 | ISBN 9781427130211 (hardcover) | ISBN 9781427130327 (softcover)
Subjects: LCSH: Desert ecology—Juvenile literature. | LCSH: Food chains (Ecology)—Juvenile literature.
Classification: LCC QH541.5.D4 W35 2021 | DDC j577.54/16—dc23

Library of Congress Cataloging-in-Publication Data

Names: Walker, Alan, 1963- author.
Title: Food chain in a desert / Alan Walker.
Description: New York : Crabtree Publishing, 2021. | Series: My first science books : a Crabtree seedlings book | Includes index.
Identifiers: LCCN 2020050997 | ISBN 9781427130211 (hardcover) | ISBN 9781427130327 (paperback)
Subjects: LCSH: Forest ecology--Juvenile literature. | Food chains (Ecology)--Juvenile literature.
Classification: LCC QH541.5.D4 W35 2021 | DDC 577.54--dc23
LC record available at https://lccn.loc.gov/2020050997

Crabtree Publishing Company
www.crabtreebooks.com 1–800–387–7650

e-book ISBN 978-1-947632-23-3
e-pub ISBN 978-1-947632-53-0

Print book version produced jointly with Blue Door Education in 2021

Author: Alan Walker
Production coordinator and Prepress technician: Tammy McGarr
Print coordinator: Katherine Berti

Printed in the U.S.A./012021/CG20201112

Photo credits: Cover page 2-3 © Ste Lane; page 4-5 © kojihirano.; page 6-7 © yhelfman; page 8-9 © By Rachel Portwood; page 10-11 © Riegsecker; page 12-13 and page 14-15 © Martin Froyda; page 16-17 © Maria Jeffs; page 18-19 © Vladislav T. Jirousek; page 20-21 plants © NoPainNoGain, hare © Anna Filippenok, snake © GoodStudio, coyote © Panaiotidi, vulture © Svetlana Foote, notebook icon © Pro Symbols All photos and illustrations from Shutterstock.com

Published in Canada
Crabtree Publishing
616 Welland Ave.
St. Catharines, Ontario
L2M 5V6

Published in the United States
Crabtree Publishing
347 Fifth Ave.
Suite 1402-145
New York, NY 10016

Published in the United Kingdom
Crabtree Publishing
Maritime House
Basin Road North, Hove
BN41 1WR

Published in Australia
Crabtree Publishing
Unit 3 – 5 Currumbin Court
Capalaba
QLD 4157